Nonroutine Problems:

Doing Mathematics

Robert London

Janson Publications, Inc.   Providence, Rhode Island

To the memory of my father, Herbert London.

He lived his life as he believed and taught me

to do the same. I am forever grateful.

98 97 96 95 94 93 92 91 90 89   8 7 6 5 4 3 2 1

ISBN: 0-939765-30-6

Contents

Acknowledgments

This book would not have been possible if it were not for the support, help, and ideas of many colleagues, friends, and students. Here I wish to acknowledge my gratitude to all those people and especially to a few who have been most directly involved with this project.

I am particularly grateful for the support of Bob Rosenbaum, the director and primary force in the PIMMS Mathematics and Science Fellows Program in Connecticut. He created and continues to create the opportunities for me and many others to grow professionally in a supportive and productive atmosphere. The ideas for and energy to implement the original nonroutine problems came out of my first summer's work as a PIMMS Fellow. The two problems on infinity developed from interest he created in the topic in his History of Mathematics class. In addition, Cliff Sloyer's classes, seminars with guest speakers, and discussions with other PIMMS fellows during the three summers I was involved full time with the PIMMS Program were the inspiration or sounding board for many of the ideas in this book.

Steve Leinwand and Bill Masalski were also instrumental in helping me select this particular project. Bill was my doctoral advisor at the University of Massachusetts, and Steve is the Mathematics consultant in Connecticut. I am also grateful to two colleagues, Frank Corbo, a PIMMS Mathematics Fellow, and John Hennelly, Language Arts Department Chairperson at Old Saybrook High School, whose comments on the manuscript were particularly helpful.

The students in my calculus classes at Old Saybrook High School beginning with the 1984-1985 school year most influenced the actual contents of this book. Their work, comments, and questions have helped determine what was necessary to create an effective curriculum of nonroutine problems.

My wife Janet and my daughter Jennifer, who wants the role of the parabola when the book is made into a movie, have given day-to-day support throughout this project. Also, it has been a pleasure working with Janson Publications, particularly Barbara Janson and Ceri Dean. Their encouragement and help have certainly influenced the quality of the book.

Introduction

In recent years the professionally aware mathematics teacher has been flooded with ideas, techniques, and materials to use in the classroom to improve problem-solving skills. Generally these materials and methods either focus on improving specific problem-solving strategies like guess and check or discover a pattern, or they require the application of a variety of straegies or a general process for solving a set of unsorted word problems. This book is concerned with a type of problem not typically included in even these improved materials: open-ended, nonroutine problems which for our purposes have the following characteristics:

1. The problems allow for various solutions.

2. The problems require the student to evaluate potential approaches to solving the problem and to select one or more to pursue.

3. The problems are such that every student with at least two years of college preparatory mathematics is able to confront the problem and generate a solution consistent with his or her ability and effort.

4. The problems generally require at least a few hours work over approximately a two-week period.

A nonroutine problem in mathematics corresponds to a creative writing assignment in language arts. Both activities require a significant investment of time and the use of higher-order thinking skills while allowing every student the opportunity to complete the assignment. Just as writing an essay can be considered the step beyond writing sentences and paragraphs, solving nonroutine problems can be considered the step beyond solving unsorted word problems or problems involving only one strategy.

Why are open-ended, nonroutine problems appropriate for inclusion in the mathematics curriculum? First, each nonroutine problem gives the students practice with three important steps of higher-order problem solving which are typically not covered or emphasized in the traditional mathematics curriculum: (1) recognizing that a difficulty exists, (2) approaching a difficult or ambiguous problem by trying something or generating data, and (3) persisting until reaching a satisfactory solution. Second, they provide an opportunity for practicing other problem-solving skills important in mathematics. These skills include: (1) finding patterns and generalizing, (2) developing algorithms or procedures and describing them, (3) manipulating symbols and numbers, and (4) reducing the problem to an easier equivalent problem.

Finally, the best reason for the inclusion of a sequence of nonroutine problems in the curriculum is the effect on the mathematical maturity of the students. It is as if the students are transformed mathematically! Instead of stopping when an obstacle is encountered, students will persist. Instead of ignoring obvious contradictions or inaccuracies, students will examine them. Instead of being intimidated by ambiguity, students will tolerate it. Instead of being satisfied with the first solution to a problem, students will work on a problem until a more satisfactory solution is reached. Instead of staring at a problem that seems unsolvable or confusing, students will try something until the problem becomes clearer. These behaviors are indicative of the gestalt of a mature mathematician. The students may still lack the experience and knowledge of the mathematician and may still make errors typical of students, but they do begin to act like mathematicians.

Survey of the Problems

The chart below gives a comprehensive overview of the problem-solving skills required in working each of the ten problems. A brief summary of each problem follows.

Nonroutine Problem	Pattern Recognition	Algorithmic Development	Manipulative Skills	Problem Reduction
1. Area, Part 1			•	•
2. $F(x)$	•		•	
3. Pi		•	•	
4. $G(x)$	•		•	
5. Area, Part 2	•		•	•
6. Infinity, Part 1	•	•		
7. Grading $G(x)$		•		•
8. n Points		•		•
9. $R(x)$	•			•
10. Infinity, Part 2	•			•

1. Calculating Area, Part 1

Students are given two areas to calculate: (1) the area within a closed irregular curve drawn on 1/4" graph paper and (2) the area bounded by $y = x^2 + 2$, $x = 0$, $x = 3$, and $y = 0$. It is assumed, of course, that students have no knowledge of how to use calculus to evaluate $\int_0^3 (x^2 + 2)\, dx$.

2. $F(x)$: Functions, Part 1

Students are given an abstract property of a function $(F(a \cdot b) = F(a) \cdot F(b))$ and three values of the function, $F(2)$, $F(3)$, and $F(5)$. They are asked to determine the values of the function for 1, 2, 3, 4, 5, 6, 7, 8, 9, 10, 11, 12, 13, 14, 15, 16, 17, 18, 19, 20. Some values can be determined directly from the given information (e.g., $F(6) = F(3 \cdot 2) = F(3) \cdot F(2)$) but six values cannot be so determined (e.g., $F(7)$); therefore, the students must develop techniques to approximate the values.

3. Pi

Students approximate the value of pi using ten different methods including a method which does not use polygons, a method which does not use trigonometry tables, a method without trigonometry, and a method without calculators. The project is graded on variety, creativity, and accuracy.

4. $G(x)$: Functions, Part 2

Students are given an abstract property of a function $(G(a \cdot b) = G(a) \cdot G(b))$ and three values of the function, $G(8)$, $G(14)$, and $G(6)$. They are asked to determine the values of the function for 1, 2, 3, 4, 5, 6, 7, 8, 9, 10, 11, 12, 13, 14, 15, 16, 17, 18, 19, 20. Students are required to develop techniques to approximate the values which cannot be determined directly from the given values.

5. Calculating Area, Part 2

Students figure out the value of t such that the area bounded by $y = 1/x$, $y = 0$, $x = 1$, and $x = t$ is equal to 1 square unit. In addition, they find t such that the area is 2 and 1000. ($t = e$, e^2 and e^{1000}.)

6. Infinity, Part 1

Students compare the size of fourteen sets including ten infinite sets. The restrictions of the problem force the students to discriminate between the size of the various infinite sets and to justify their answers. The sets include $\{1, 2, 3, \ldots\}$, $\{1000, 2000, 3000, \ldots\}$, {all real numbers between 0 and 1}, and {all functions from $(0, 1)$ onto $(0, 1)$}.

7. Grading $G(x)$

Students develop a short yet valid method for grading the results of nonroutine problem 4 ($G(x)$: Functions, Part 2).

8. Connecting n Random Points

Students develop a procedure to connect n random points so that the total distance is minimized.

9. $R(x)$: Functions, Part 3

Students are given an abstract property of a recursive function $R(x) = R(x - 1) + x$ and $R(1) = 2$, and asked to determine (and justify) a variety of values of the function such as $R(-1000)$, $R(55\frac{1}{2})$, and $R(8.732)$.

10. Infinity, Part 2

Students simplify eleven expressions involving one or more of the first three transfinite numbers, e.g., $\aleph_0^2$, $2^{\aleph_0}$, and $2^{2^{\aleph_0}}$.

General Pedagogical Notes

In order to achieve the implied objectives, students must complete a sequence of eight or ten nonroutine problems over the course of a year. Here we address general issues in the implementation of nonroutine problems. Detailed instructions for each individual problem, including descriptions of sample solutions and enrichment/extension topics, are given in the body of the text.

Structuring the Assignment

When giving directions for each problem, it is important to supply enough information so that the problem will be challenging, yet workable. The problem should not be straightforward, but each student should be able to work on it without unproductive frustration. Generally, students who have completed two years of academic mathematics, *e.g.*, Algebra 1 and Geometry, have the prerequisite skills to benefit from these nonroutine problems.

It should be made clear to students what is expected: namely, that the problems require work evenly spread over about two weeks, and that their write-up should include not only their solution but also strategies they attempted but discarded, their rationale for selecting the final solution, and enough detail to support their work.

Encourage students to work at least some on the problem the first night, emphasizing that this allows their ideas to develop, even when not directly working on the problem. Although solutions should be primarily individual efforts, they are encouraged to use available resources. For example, the teacher can serve as a sounding board for ideas, giving feedback or hints which will allow students to work productively. Other outside references and resources which could be used include computers, science equipment, and mathematics texts from previous courses. Of course, a distinction is made between using a resource as an aid in solving a problem and using a resource to solve a problem — it is inappropriate to ask a mathematician for a solution or to try to look up a solution in a book. Occasional experience working with another student is very valuable and probably should be required for at least two problems.

In order to prevent students from saving the assignment until the last weekend, you may require them to turn in a progress report, summarizing work to date, after the first week. This report will not only indicate what the student has done but also will allow an opportunity to determine which students need assistance or hints.

The order in which the ten problems are presented in this work, while not the only effective order, is based on the following observations: (1) Experience shows that the problem Calculating Area, Part 1 is an excellent first problem. (2) The problems Infinity, Part 1, Infinity, Part 2, and Connecting *n* Random Points are more appropriate as later problems because of their difficulty levels. (3) The three problems on functions are less difficult and can help build student confidence at key points. (4) Related problems, the two area problems, the three function problems, and the two infinity problems, seem to be most effective when at least one nonrelated problem is completed between them.

Class Discussion

To help students during the discussion of their work and the problems, the teacher needs to be familiar with the three steps of problem solving (recognizing that a difficulty exists, trying something, and persisting) and the four problem-solving skills (pattern recognition, algorithmic development, manipulative skills and problem reduction). When discussing individual problems, the teacher should focus attention on these steps and skills whenever possible, especially when a student's solution is an example of how to use these steps or skills. Patiently contrasting different student solutions or contrasting teacher-prepared solutions with student solutions is quite beneficial over the course of a school year.

Discussion of solutions is very important and needs to be carefully prepared. To be valuable, it needs to respond to how your students approached the problem. Since the nonroutine problems require a level of skill and effort not normally required of students in mathematics, we expect that students will initially experience difficulty and that their solutions will seem inadequate. The quality of solutions needs to be a central focus of the discussion (and directions) for the first few problems. Suggestions for focusing on the quality of solutions have been included with the first few problems. In summary, what is needed is a supportive atmosphere for gradual growth.

In addition to emphasizing the already mentioned problem-solving skills, the teacher can use the discussion to draw attention to well-documented solutions, the clever use of resources, or unusual approaches. The discussion also can be an opportunity to examine common errors that students make, such as not persisting long enough with a problem, generalizing with inadequate data, being careless, and not exploring conflicting data or obvious errors.

After the initial focus on students' solutions, it is appropriate to discuss classical solutions or the historical significance of the problem. A suggested outline for this part of the discussion has been included with several sets of teaching suggestions.

Evaluation

To be effective, the nonroutine problems should be seen as an integral part of the grading process for the course. However, the experience of working on the nonroutine problems should be a positive experience which suggests that unnecessary concern for the grade that will be given would be inappropriate. This implies a grading system like the following: 90 – 100% means excellent work, indicating insight and/or effort beyond what one would normally expect; 80 – 89% means good to very good work, indicating that the student spent a good amount of time on the problem, documented the work and gained reasonable insight into the problem; 70 – 79% means fair work, indicating that the student put in at least the minimal acceptable effort but clearly could have gone at least a step further; and unacceptable, meaning that the student clearly did not spend enough time on the problem to learn anything significant.

Problem 1

Calculating Area, Part 1

Find the two areas given here.

A. The area of the closed irregular curve drawn below.

B. The area bounded by $y = x^2 + 2$, $x = 0$, $x = 3$, and $y = 0$.

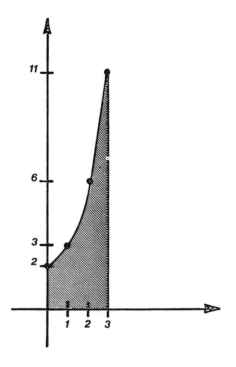

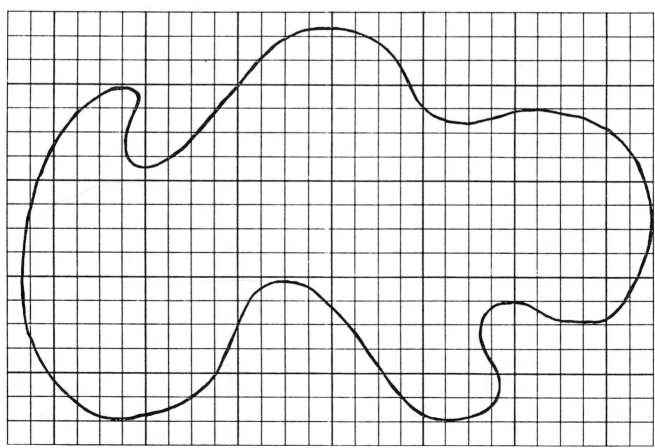

Extension and Enrichment

1.

Part A lends itself to an exploration of Monte Carlo methods using a computer. The program below prints 30 pages, each page containing 126 random points (a, b), a between 0 and 28 and b between 0 and 18. Let x equal the area within the irregular figure. Estimate the area of the irregular figure as follows.

$$\frac{x}{\text{area of grid}} = \frac{\text{no. random points within figure}}{\text{total no. points}}$$

$$\frac{x}{(4 \times 126)} = \frac{\text{points within figure}}{126}$$

$$x = 4 \text{ (no. of points within figure)}$$

With 30 pages, each student can estimate the area and results can be compared. Since accurate counting is important, students may want to check each other's work. This process results in an estimate for the area of the irregular figure of between 265 and 270 units.

```
10 PR#1
20 FOR Y = 1 TO 30
30 FOR X = 1 TO 126
40 A = (RND(1)*281)/10
50 B = (RND(1)*181)/10
60 PRINT A;", ";B,
70 IF X/2 = INT(X/2) THEN PRINT
80 NEXT X
90  PRINT:PRINT:PRINT:PRINT:PRINT
100 NEXT Y
110 PR#0
```

2.

Part B lends itself to an exploration of the trapezoidal method of calculus. This program calculates the area under the parabola using the trapezoidal method, with n trapezoids, for n input by the user. Students can experiment with different values of n, perhaps with the goal of discovering what value of n will result in an area of 15 square units, plus or minus a specified error.

```
10 PRINT "HOW MANY INTERVALS?"
20 INPUT N
30 A = 0:B = 3/N:S = 0:T = 0
40 FOR X = 1 TO N
50 H = 3/N
60 B1 = A^2 + 2
70 B2 = B^2 + 2
80 T = .5*(B1 + B2)*H
90 S = S + T
100 A = A + 3/N
110 B = B + 3/N
120 NEXT X
130 PRINT "THE AREA IS ";S
```

3.

Another interesting technique for finding the area is to use mean ordinates. For example, in the figure, one can estimate the position of a horizontal segment TW so that the areas of the two "triangular" regions RTV and SVW are nearly equal. Then the area of rectangle $PQWT$ is approximately equal to the area under the arc RS, $i.e.$, $\bar{y} \times PQ$ gives the desired area. $\bar{y}$ is called the mean ordinate or height of the curve in the interval PQ.

It is interesting for students to compare this method to the trapezoidal method, using the area of trapezoid $PQSR$, and to determine which is more accurate. If a reasonable scale is used to enlarge the figure, the mean-ordinate method should be more accurate.

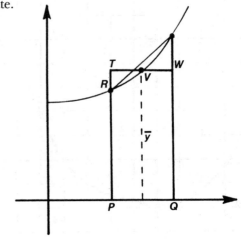

Teaching Suggestions

1.

When you distribute the first problem, you might want to clarify the assignment by charging students with finding the area as accurately as they can, using the following illustration: Assume that your employer is competing for a job with several other companies. The company whose solutions for the two areas are nearest the actual areas will receive the job.

2.

It is useful for students to realize the following.

a. In the first problem, each 1/4" square represents one square unit.

b. In the second problem, the area can be calculated fairly easily using calculus, but that is not acceptable for this assignment.

c. In the second problem, each point on the curve can be determined, *i.e.*, $(x, x^2 + 2)$.

3.

Because this is the first nonroutine problem, it may be wise to spend some time discussing expectations. For example, you might want to emphasize points discussed in the pedagogical notes concerning the time required to produce a solution, documentation requirements, the existence of many possible solutions, and the possibility of refining solutions.

4.

Experience indicates that most students do not have the experiences and skills necessary to answer this first nonroutine problem at the level required for later problems. You could give students hints and information, but it is more productive in the long run to give the directions as outlined and then use the discussion of solutions as an opportunity for the students to contrast their solutions with "higher quality" solutions. Suggestions for how to do this are included with the solutions.

5.

Because students lack experience with nonroutine problems, you may wish to evaluate the first assignment differently from succeeding ones. For example you might assign one grade based on anticipated later standards and one for this problem, scaled upward. This procedure allows students to recognize their baseline skills, but does not penalize them for their initial lack of skill. For example, an area between 265 and 270 square units is better than average for the first problem, and an area between 14.75 and 15.25 square units is better than average for the second problem.

Student Solutions

1.

Many student solutions for this problem amount to counting the fully enclosed squares and estimating the remaining ones to be either 1/2 or 1 square unit. Here is a typical example.

"I began with the counting of each . . . complete box within the boundaries of the figure . . . Next I tracked down every 1/2 box . . . Next I used a matching system to further pursue my final sum. Looking for [two boxes] that would together make one, I discovered a total of 28 complete boxes, leaving one box [which] I determined to be 3/4 of a square unit and . . . I came up with a grand total of $282\frac{3}{4}$ square units, the area for the entire figure."

2.

For part B, many students calculate the area to be $15\frac{1}{2}$ square units by "replacing" the curve $y = x^2 + 2$ with the straight lines connecting (0,2) and (1,3), (1,3) and (2,6), and (2,6) and (3,11) and then dividing the area into three trapezoids or, more likely, three rectangles and three triangles as shown here.

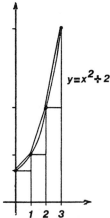

An important objective in discussing this assignment is to help students recognize the difference between the above solutions and those of higher quality. This is somewhat delicate; you need simultaneously to help students see the inadequacy of their solutions yet reassure them of their potential to generate higher-quality solutions.

The second problem presents an excellent opportunity for this type of discussion. Specifically, an obvious improvement to the solution cited above is to increase the number of trapezoids used. A discussion of this issue gives students an example of a refinement of their solution that is similar enough to reassure them of their potential for future success.

3.

Here is an example of a good approach to the problem:

"When I first looked at this problem, I didn't think I could solve it. My first thought was to count the squares, but when I actually started to count them, I found out how difficult and inaccurate it would be. I thought about breaking it down into other shapes but with the inconsistent curves, it seemed impossible. Then I came up with my final solution."

This example not only reinforces the quoted approach, but also helps other students see that they are capable of this type of reasoning. Notice that the quotation well illustrates the three steps of problem solving: recognizing that there is a difficulty, trying something, and persistence.

4.

The following samples for the irregular figure could also be discussed.

a.   One student first used estimates of just halves and wholes, then tried estimates into fourths, and finally used blocks divided into sixths.

b.   Another student first totalled all the squares that were exactly or very close to either 1 or 1/2 square unit. To get a good estimate of the remaining area, the student then cut out all the remaining partial squares and fit them into a rectangle on graph paper.

c. Some students used weighing as a technique. For example, the entire rectangle and then the irregular figure were weighed and a proportion was set up to solve for the unknown area. A somewhat less accurate variation involves weighing one square unit and the irregular figure.

d. Another calculated the average width and length, then multiplied to calculate the area.

e. Quarter or half circles were used to approximate portions of the figure.

f. Some students calculated two areas: the area within the irregular figure and the area outside the irregular figure but within the rectangle. Then, as a check for their method, they added the two areas and compared the result to the actual total area of the rectangle.

5.

Other solutions for the second problem include enlarging the graph or using parabolas or circles to estimate the area. One student wrote a computer program which used the trapezoid method to calculate the area. The student increased the number of trapezoids until it seemed clear that the area approached 15 square units. As a check, a second program was written to calculate the area above the curve and below $y = 11$. The two areas should total 33 square units.

6.

Student errors in reasoning can also be useful as you introduce students to nonroutine problems. For example, one student wrote:

"My first solution is to calculate the area outside of the figure [by finding] the average length of all the lines outside the figure. The average was 7.8. Next, I multiplied that number by the width of the graph which was 28 [for] a total of 218.4 . . .I then subtracted this from the total area . . . which was 504. I came up with an answer of 313.6."

"My second solution is to find the average distance of the lines vertically inside the figure [10.6] . . . multiplied by the width inside the figure which was 27 . . . for an area of 286.2."

Ask students if they notice anything unusual about the solutions. They should notice that there is a substantial difference in the two calculations of the same area. Ignoring an obvious inconsistency is not unusual, especially in beginning problems. Of course, it should be noted that the student's error was simply subtraction.

$F(x)$: Functions, Part I

I.

You are given the following information about a function $F(x)$:

A. For $a, b, F(a \cdot b) = F(a) \cdot F(b)$

B. $F(2) \approx 2.585$

   $F(3) \approx 4.505$

   $F(5) \approx 9.070$

As best you can, complete the following values, rounding your answer to the nearest thousandth:

$F(1) =$	$F(8) =$	$F(15) =$
$F(2) = 2.585$	$F(9) =$	$F(16) =$
$F(3) = 4.505$	$F(10) =$	$F(17) =$
$F(4) =$	$F(11) =$	$F(18) =$
$F(5) = 9.070$	$F(12) =$	$F(19) =$
$F(6) =$	$F(13) =$	$F(20) =$
$F(7) =$	$F(14) =$	

You will probably not be able to determine some values to the nearest thousandth directly from the given information. Part of the assignment is to determine your best approximation for these values.

II.

On separate paper document and explain the strategies you used and how you determined each of your values.

Extension and Enrichment

1.

One natural extension for this problem is a discussion of why a number raised to the zero power is equal to one.

Although the idea of raising something to the zero power does not make intuitive sense, students can see that inductively it does make sense by looking at the following patterns:

$$5^4 \ = \ 625$$
$$5^3 \ = \ 125$$
$$5^2 \ = \ 25$$
$$5^1 \ = \ 5$$
$$5^0 \ = \ ?$$

In other words, the definition is consistent with the pattern of exponents. Similarly, one can see that the definition is consistent with the law of exponents: $x^a/x^b = x^{a-b}$ (x not equal to 0) by noticing that $1 = x^a/x^a = x^{a-a} = x^0$.

Equally interesting subjects for discussion include raising numbers to imaginary powers (discussed in "Calculating Area, Part 2"), irrational powers or transcendental powers such as pi.

2.

Notice that $F(x)$ is a nonintegral rational root. An interesting historical note is that numbers which we consider commonplace, such as the square root of two, were not discovered until the time of Pythagoras. Using an indirect proof, he showed that the length of the hypotenuse of an isosceles right triangle with legs of one unit was not a rational number. He assumed that the hypotenuse h was rational *i.e.*, $h = p/q$ or $h^2 = 2 = (p/q)^2$, where p and q are integers. Assuming that p/q is in reduced form, he argued that p and q could not both be even (if so,

p/q could be reduced by dividing p and q by 2). The equation $2 = (p/q)^2$ can be rewritten: $2q^2 = p^2$. He then looked at the three remaining possibilities for p and q. The first possibility is that p and q are both odd; this is not possible because $2q^2$ would be even and p^2 would be odd. The second possibility is that p is odd and q is even; this is not possible because $2q^2$ would be even and p^2 would be odd. The third possibility is that p is even and q is odd; this is not possible because if p is even then $p = 2r$ for some r, so we can rewrite our equation to get $2q^2 = 4r^2$; dividing each side by 2 we get $q^2 = 2r^2$ — a contradiction since q^2 is odd and $2r^2$ is even. By eliminating all four possibilities we have contradicted the original assumption thereby proving that the square root of 2 is not rational.

After demonstrating the first three cases, the fourth case and the remainder of the proof may be assigned to the students.

3.

Two of the remaining nonroutine problems concern functions defined by an abstract property: $G(ab) = G(a) + G(b)$ and $R(x) = R(x-1) + x$. A function with the property that $F(a+b) = F(a) + F(b)$ is somewhat easier to study and could be useful as an extension or introductory activity for this lesson. Notice that $F(0) = 0$; also if $F(1) = m$ then $F(n) = mn$.

Teaching Suggestions

1.

To insure that the students understand the meaning of the abstract property of $F(x)$, review how to calculate $F(6)$: $F(6) = F(3 \cdot 2) = F(3) \cdot F(2) = 4.505 \cdot 2.585 = 11.645$.

2.

Explain that some of the values (*e.g.*, $F(13)$, $F(17)$,...) cannot be determined by using only the given values. Tell the students that the function is not linear and that for this assignment it is *not* acceptable to determine an unknown value by simply averaging known values (*i.e.*, $F(17) \neq (F(16) + F(18))/2$). Remind students that part of the assignment is to document their work, including all nontrivial strategies and justification for the strategy selected.

3.

You may find that the quality of many of the student solutions is not significantly better than for the first problem. To help students make the transition to better quality solutions, you may want first to review their work, provide appropriate feedback, and allow them extra time for improving inadequate solutions.

Student Solutions

1.

The following are straightforward calculations that all students should get: $F(4) = F(2) \cdot F(2)$, $F(6) = F(2) \cdot F(3)$, $F(8) = F(2) \cdot F(2) \cdot F(2)$, $F(9) = F(3) \cdot F(3)$, $F(10) = F(2) \cdot F(5)$, $F(12) = F(2) \cdot F(2) \cdot F(3)$, $F(15) = F(5) \cdot F(3)$, $F(16) = F(4) \cdot F(4)$, $F(18) = F(2) \cdot F(9)$, $F(20) = F(4) \cdot F(5)$. In addition, many students realize that $F(1)$ can be calculated by noticing that $F(a) = F(a \cdot 1) = F(a) \cdot F(1)$; therefore, $F(1) = 1$ (assuming $F(x)$ not equal to 0). The heart of the problem is to determine good approximations for $F(7)$, $F(11)$, $F(13)$, $F(14)$, $F(17)$ and $F(19)$.

The most common strategy for determining the remaining values is to attempt to discover a pattern of how the values increase. For example, by graphing the given and computed values carefully on a Cartesian coordinate system one can sketch a smooth graph of $y = F(x)$ and thereby approximate the intermediate values. Or by studying the values of $F(x+1)/F(x)$ for $x = 1$ to 20 (when possible), one can approximate some unknown values. Some students include values of $F(x)$ for fractional values of x such as $3/2$, thereby increasing the accuracy of their approximations.

A more sophisticated solution is to notice that the graph is similar to that of a parabola and to use three given values to determine a parabola (or parabolas) that approximates the actual graph. Some students mistakenly become convinced that the graph of the function *is* a parabola (vs. using parabolas which approximate the curve) and are obviously unsuccessful in finding a parabola that works. This error and other unsuccessful strategies lead some students to conjecture that $F(x) = x^n$. They then use a calculator to determine the value of n.

Other students use values of $x > 20$ to approximate unknown values; *e.g.*, $F(121) = 2F(11)$ and $F(120) = F(10)F(12)$, so $F(11) \approx [F(10) \cdot F(12)]/2$.

2.

Because the given values are approximate (rounded off to the nearest thousandth), there can be some discrepancy in the "best" approximations. For example, using the function $F(x) = x^{1.37}$ you get slightly different values than if you use the abstract property and the given values. The primary values given below were determined using $F(x) = x^{1.37}$; the values in parentheses were determined directly from the abstract definition and the given values.

$F(1) = 1.000$	$F(11) = 26.712$
$F(2) = 2.585$	$F(12) = 30.094 \ (30.102)$
$F(3) = 4.505$	$F(13) = 33.582$
$F(4) = 6.681 \ (6.682)$	$F(14) = 37.170$
$F(5) = 9.070$	$F(15) = 40.855 \ (40.860)$
$F(6) = 11.643 \ (11.645)$	$F(16) = 44.632 \ (44.649)$
$F(7) = 14.381$	$F(17) = 48.497$
$F(8) = 17.268 \ (17.273)$	$F(18) = 52.447 \ (52.461)$
$F(9) = 20.291 \ (20.295)$	$F(19) = 56.480$
$F(10) = 23.442 \ (23.446)$	$F(20) = 60.591 \ (60.606)$

3.

One good focus for the discussion is the strategy of looking for patterns. The chart below is one student's attempt to organize the data to find patterns. The formulas for each heading have been added. The values calculated by the student are in brackets.

$F(x)$					
How Determined	Function	Values	% of Next Value	% Increase	Numerical/ Value Increase
	$f(1)$	[1.1374]	[44%]		
Given	$f(2)$	2.585	57.38%		[+1.45]
Given	$f(3)$	4.5050	67.42%	10.40%	1.92
$f(2) \cdot f(2)$	$f(4)$	6.6820	73.67%	6.25%	2.177
Given	$f(5)$	9.0700	77.89%	4.22%	2.388
$f(2) \cdot f(3)$	$f(6)$	11.6450	[81.04%]	[3.15%]	2.575
added 2.724	$f(7)$	[14.359]	[83.19%]	[+2.15%]	[+2.724]
$f(2) \cdot f(4)$	$f(8)$	17.2730	85.11%	1.92%	[+2.094]
$f(3) \cdot f(3)$	$f(9)$	20.2950	86.56%	1.50%	3.022
$f(2) \cdot f(5)$	$f(10)$	23.4460	87.99%	1.47%	3.151
added 3.2	$f(11)$	[26.647]	[88.52%]	[ ]	[3.2]
$f(2) \cdot f(6)$	$f(12)$	30.1020	[89.13%]	[+11.14%]	[+3.456]
added 3.094	$f(13)$	33.1960	[90.06%]	[+.93%]	[+3.094]
added 3.335	$f(14)$	[36.546]	[90.83%]	[+.77%]	[3.35]
$f(3) \cdot f(5)$	$f(15)$	40.8600	91.51%	[+.68%]	[+4.314]
$f(4) \cdot f(4)$	$f(16)$	44.6490	92.16%	0.65%	3.789
added 3.8	$f(17)$	[48.499]	92.35%	0.19%	3.8
$f(3) \lozenge f(6)$	$f(18)$	52.4615	[92.92%]	[+.76%]	4.016
added 4	$f(19)$	[56.460]	[93.16%	[+.24%]	[+4]
$f(4) \cdot f(5)$ or $f(2) \cdot f(10)$	$f(20)$	60.6060			[+4.146]

The reader might notice the obvious discrepancy in the last column — the numerical value increases for $F(13)$ and $F(14)$ are less than some previous increases and therefore do not fit the pattern. In the margin next to these values the student wrote: "the increase is stopped here, can't understand."

This example provides another focus for the discussion period — persistence. Despite the excellent effort required to create the chart, the student did not experiment with adjusting some of the values so that the entire chart was consistent.

4.

Many students are unable to accurately judge the adequacy of their solutions; therefore, they tend not to persist for the beginning problems. For example, some weaker solutions use graphing for their final solution to obtain values between .1 and .2 off the actual answers (compared to less than .02 for better solutions). Other students may start with graphing but recognize its inadequacy and persist until finding the function.

Approximating Pi

For this assignment you are to approximate pi using ten different methods. For each method you must include your data and describe the instruments you used. Five of these methods must be labelled: (1) a method which does not involve polygons, (2) a method which uses trigonometry but does not use trigonometry tables or a calculator, (3) a method without trigonometry, (4) a method without a calculator and (5) your best method.

Your grade will be based on the total number of points you score. You will be given 0 to 20 points for each of the following categories: (1) your best method, (2) your other four labelled methods, (3) originality and creativity, (4) variety of your methods and (5) the quality and accuracy of your calculations.

Extension and Enrichment

1.

Perhaps the most famous approximation of pi was by Archimedes in the book *The Measurement of a Circle* in which he makes the proposition that "The circumference of any circle exceeds three times its diameter by a part which is less than $\frac{1}{7}$ but more than $\frac{10}{71}$ of the diameter", *i.e.* $3\frac{1}{7} > \pi > 3\frac{10}{71}$. This approximation is within 0.2 percent of the actual value.

2.

The technique that Archimedes used is of interest because it relates to two other nonroutine problems. Firstly, he used methods similar to the methods used for the area nonroutine problems. Specifically, he inscribed and circumscribed regular polygons with sides of 6, 12, 24, 48 and 96 sides. A hexagon is simple to inscribe by marking off 6 radii to locate the vertices. After constructing the tangents to the vertices, you can construct the circumscribed hexagon. Then by bisecting subtended arcs the other polygons can be constructed.

Secondly, he used a recursive function to solve for pi; namely:

$$P_{2n} = \frac{2 p_n P_n}{p_n + P_n} \qquad p_{2n} = \sqrt{p_n P_n}$$

where P_n = perimeter of the inscribed polygon with n sides and p_n = perimeter of the circumscribed polygon with n sides. He knew that $P_6 = 3d$ and $p_6 = 2\sqrt{3}$, and that $265/153 < \sqrt{3} < 1351/780$. The practical limits of the mathematics of Archimedes's time made calculations beyond a polygon of 96 sides prohibitive.

3.

The extension and enrichment section of "Area, Part 2" discusses the meaning of $e^{i\pi} + 1 = 0$.

Teaching Suggestions

1.

This problem emphasizes evaluative skills. The student must decide on ten methods which together will result in the highest possible grade on the project.

2.

To introduce the assignment and to stimulate thinking about various types of solutions, you might give the students an example of approximating pi such as the following:

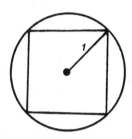

Since the radius of the circle is equal to one, the side of the square is equal to $\sqrt{2}$, the circumference of the circle is equal to 2π, and the perimeter of the square is equal to $4\sqrt{2}$. If we say that the circumference of the circle is approximately equal to the perimeter of the square, then we have the following:

$$2\pi \approx 4\sqrt{2}$$

If we substitute 1.414 for the square root of two and solve for pi we get that pi is approximately 2.828.

3.

Students may use formulas involving circumference, area, volume or physical forces (such as centrifugal force), but not infinite series such as $\pi/4 = 1 - 1/3 + 1/5 - 1/7 + \cdots$ or trigonometry formulas such as $\pi/4 = 4 \arctan(1/5) - \arctan(1/239)$. You might encourage students to ask you individually about the appropriateness of a questionable method.

4.

The rationale for the above directions is to increase the likelihood that the student will use a variety of methods. Consequently, in discussing the directions, it could be useful to point out that the method used above to approximate pi ($\pi = 2.828$) is not very accurate and that the method could be improved by increasing the number of sides (you might want to give an additional example). At the same time, make it clear that a set of solutions in which most of the methods involved polygons (even good approximations) would not receive a good grade, certainly losing points for originality and variety. Additionally, students should realize that the five categories for grading force them to evaluate the ten methods they include for balance — a very creative method may not be very accurate while an accurate method may not be very creative, etc. By emphasizing both accuracy and creativity, you will discourage the students from "fudging" their data.

5.

For this problem, it would be particularly appropriate to have the students work in pairs and to check each group's progress after the first week.

Student Solutions

1.

The work of two students will be contrasted to give a sense of the range of solutions for this problem. The first student received a fair grade and the second student received an excellent grade. For the best method the first student used the law of cosines with a 360-sided polygon inscribed in a circle to obtain a value of 3.1415513; the second student wrote a program for calculating the area of a quarter of the unit circle by rectangular approximation. Using 100,000 rectangles the student got a value of 3.141592682 (the first eight digits are accurate).

For the four other labelled methods the first student: (a) compared an inscribed circle with a square by weighing for a value of 3.08 (without trigonometry), (b) measured the circumference by rolling a circle on a ruler and solved for pi for a value of 3.14 (without a polygon), (c) used the law of cosines with a 45 degree angle for a value of 3.06 (trigonometry without charts) and (d) used an inscribed hexagon for a value of 3 (without calculator). The second student: (a) used a cylinder, density formula, calipers and three trials for a value of 3.11 (without trigonometry), (b) used a sphere and determined volume by displacement for a value of 3.1746 (without a polygon), (c) used an inscribed hexagon for a value of 3 (trigonometry without charts) and (d) wrapped string around a circle ten times and used a caliper to measure the diameter for a value of 3.136 (without calculator).

For the five remaining methods the first student: (a) used a pendulum ($T = \pi$ [square root of l/g]) for a value of 3.1474185, (b) used the law of cosines with an angle of 5 degrees for a value of 3.1406, (c) used a cylinder and displacement for four runs for a value of 3.13, (d) circumscribed a square around a circle for a value of 4 and (e) averaged the area of a circumscribed and inscribed triangle for a value of 3.8971. The second student: (a) used centripetal force ($W =$ the square of $[2\pi r/T]$ times m/r) for a value of 3.144418905, (b) used the fact that the

acceleration of gravity equals the acceleration of a pendulum to calculate pi as 3.14128629, (c) inscribed a polygon with 2 raised to the twelfth power sides in a circle for a value of 3.141581267, (d) circumscribed a similar polygon for a value of 3.14159327 and (e) used a program to calculate the length of an arc equal to a quarter of a circle using 100,000 line segments for a value of 3.141638452.

2.

One useful technique for discussion of this particular problem is to select two methods from each student — one which includes a positive point in reference to the problem-solving skills and one which includes an aspect which could be improved. The methods you pick for improvement might all focus on one skill, manipulative skills for example. The question for discussion could be how to improve the accuracy of the calculations. Could the quality of measurements have been increased by more trials? What instruments were used and could they have been improved? Were the calculations checked?

3.

The third step of problem solving, persistence, might be a good focus for the discussion. Students can compare methods in which they persisted and felt satisfied versus methods which they gave up on or "settled" for a less than adequate solution. Discarded methods could be discussed to see if persistence might have increased the effectiveness of the method.

4.

This problem is good for focusing on the adequacy of documentation. You might pick out the best and the worst documented method for each student either for individual or class discussion.

Problem 4

$G(x)$: Functions, Part 2

I.

You are given the following information about a function $G(x)$:

A. For all $a, b > 0$, $G(a \cdot b) = G(a) + G(b)$

B. $G(8) \approx 1.660$

   $G(14) \approx 2.107$

   $G(6) \approx 1.430$

As best you can, complete the following values:

   $G(1) =$ $G(11) =$

   $G(2) =$ $G(12) =$

   $G(3) =$ $G(13) =$

   $G(4) =$ $G(14) = 2.107$

   $G(5) =$ $G(15) =$

   $G(6) = 1.430$ $G(16) =$

   $G(7) =$ $G(17) =$

   $G(8) = 1.660$ $G(18) =$

   $G(9) =$ $G(19) =$

   $G(10) =$ $G(20) =$

II.

Explain (on a separate sheet) how you determined each of your values.

Extension and Enrichment

1.

This problem is an excellent introduction to the topic of logarithms. I like to draw the students' attention to the relatively slow growth of logarithms compared to other increasing functions such as $y = x^2$ and $y = 2^x$. This topic of growth is also relevant to the problems "Area, Part 2" and "Infinity, Part 2" (see Extension and Enrichment for those problems) and to binary sorts in computer programming.

2.

There is a fairly straightforward method for determining the base of a logarithm given just one value. For example, if $G(k) = 1$, then k is the base and k can be determined given any value of $G(x)$; e.g. since $G(8) = 1.660$ then $(1/1.660)$ $G(8) = 1$ which means that $G(8^{1.660}) = 1$ so the base is $8^{1.660} \approx 3.5$.

3.

When John Napier (1550–1617) discovered logarithms, the discovery was considered one of the greatest the world had seen. David Burton (*The History of Mathematics: An Introduction*, Allyn and Bacon, 1985; pages 326–37) writes: "Seldom has a new discovery won such universal acclaim and acceptance. With logarithms, the operations of multiplication and division can be reduced to addition and subtraction, thereby saving an immense amount of calculation, especially when large numbers are involved. Since astronomy was notorious for the time-consuming computations it imposed, the French mathematician Pierre deLaplace was later to assert that the invention of logarithms 'by shortening the labors, doubled the life of the astronomer'."

To give the students a better sense of the importance of the discovery, you could have the students do a relevant calculation without logs (and no calculators!) and then with logs.

Teaching Suggestions

1.

This assignment is similar to "*F(x)*: Functions, Part 1" but is more difficult because: (a) $G(x)$ is a log function (which most students have more difficulty understanding) and (b) the given values for $G(x)$ are more difficult to work with than the given values for $F(x)$.

2.

For "*F(x)*: Functions, Part 1" the students were shown how to calculate $F(6)$ to insure that they had enough of an understanding of the definition of $F(x)$ to successfully complete the straightforward part of the assignment. It is suggested *not* to give the students any direct help for this problem. The rationale for this suggestion is that a key component of this problem is for students to figure out how to use the given values to obtain most of the other values. For example, the students need to see that $G(8) = 3G(2)$; therefore, $G(2)$ can be calculated to be $G(8)/3$. You might, however, clarify the assignment by giving them a made up example; *e.g.* if $G(2) = 2.310$ and $G(3) = 3.257$, then $G(6) = G(3) + G(2) = 2.310 + 3.257 = 5.567$.

3.

It may seem that this problem is fairly easy since students have already completed "*F(x)*: Functions, Part 1." The average student, however, either never realizes, or does not immediately realize, that the function is a log function. Experience with "*F(x)*: Functions, Part 1" does help the student see possible strategies to solve the problem and to reach an adequate solution. Consequently, this problem not only challenges the student, but also helps improve the student's confidence in his/her ability to solve nonroutine problems.

4.

Nonroutine problem #7, "Grading $G(x)$," involves the evaluation of solutions to this $G(x)$ problem. You might want to read the directions for "Grading $G(x)$" before returning student work for this problem.

Student Solutions

1.

Overview: The key to starting this problem is to realize that $G(8) = G(2 \cdot 2 \cdot 2) = 3G(2)$, so $G(2) = G(8)/3$. Then $G(3)$ can be calculated by realizing that $G(6) = G(3) + G(2)$, so $G(3) = G(6) - G(2)$. Similarly, $G(7)$ can be calculated. Then by direct substitution one can calculate $G(4)$, $G(9)$, $G(12)$, $G(14)$, $G(16)$ and $G(18)$. Also, it can be seen that $G(1) = 0$ (*e.g.* $G(a \cdot 1) = G(a) + G(1)$; so $G(1) = 0$). As in the problem "$F(x)$: Functions, Part 1," the remaining values can be determined by discovering a pattern in how the function increases. Of course, if the student realizes that $G(x)$ must be a log function ($G(x) = \log_{3.5} x$), then it becomes fairly straightforward to determine the remaining values. For example, most students who determine that $G(x)$ is related to logarithms work with logs base 10; specifically they determine that $G(x) = (1.838) \log(x)$.

2.

Some methods that might be particularly useful for helping students understand the three steps of problem solving and the other problem-solving skills for this problem include: (a) comparing and contrasting the student solutions to this problem with their solutions to "$F(x)$: Functions, Part 1"(*e.g.* did the students demonstrate more persistence?); (b) focusing on the manipulative skills by discussing whether typical solutions (*e.g.* graphing, studying patterns of increase) could have been improved in accuracy, or contrasting two solutions that used similar methods but differed in accuracy; and (c) discussing solutions in which students discovered that $G(x)$ was a logarithm (especially if the student attempted other solutions first), focusing on what preceded the solution.

3.

It can be useful to compare solutions which reach similar conclusions by different steps. In the example that follows, both students concluded that $G(x)$ was equal to some constant times the log of x. The first student quickly realized that the problem could be solved using logarithms. After solving for the straightforward values, the student solved for k in $k \log(x) = G(x)$ by using $x = 6$ ($k = G(6)/\log 6 = 1.838$) and checked the answer for other values of x. Only the exceptional student solves this problem so directly. The second student first tried generating equations involving unknown values of $G(x)$ but always ended up with more unknowns than equations. The student then worked with fractional values, for example, $G(4/3) = G(8/6) = G(8) - G(6) = .230$, to generate a fairly good graph of $G(x)$. After realizing that $G(27) = G(3/4) + G(36)$ and $G(27) = 3G(3)$, thus $G(3) = .877$, the student calculated $G(2) = G(8)/3$ and the other straightforward values. From the graph of $G(x)$, the student conjectured that $G(x)$ was related to $\log x$ and soon discovered that $G(x) = 1.838 \log(x)$.

4.

Other strategies that students have used include: (a) estimating the value of $G(x)$ by noticing that the difference between consecutive integral values of $G(x)$ decreases as the value of x increases (for example, $G(7)$ can be estimated by calculating $G(49) = (G(48) + G(50))/2$ and noticing that $G(49) = 2G(7)$); (b) carefully graphing $G(x)$, getting within .005 of the actual values; and (c) using trial and error to "solve" a system of simultaneous equations. Although there are more unknowns than equations, a fairly good solution can be obtained by trial and error.

Problem 5

Calculating Area, Part 2

Below is a sketch of a region bounded by the graphs: $y = 1/x$, $y = 0$, $x = 1$, and $x = t$.

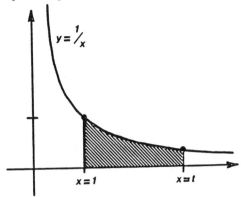

A. Find t such that the area of the region is one. Round off your answer to the nearest thousandth.

B. Find t such that the area of the region is two. Round off your answer to the nearest thousandth.

C. Find t such that the area of the region is 1000. You may write your answer as an unsimplified arithmetic expression; *e.g.* $9876 \times 4873/45$ or $86^{14} + 14^{21}$.

Extension and Enrichment

1.

The following program allows one to approximate the value of *t* for an area of one using rectangles (above the curve) and/or trapezoids:

```
10 PRINT "ENTER YOUR VALUE FOR THE CHANGE IN X"
20 INPUT DX
30 PRINT "TYPE 1 IF YOU WANT TO USE THE RECTANGLE
   METHOD, 2 OTHERWISE"
40 INPUT S
50 IF S <> 1 THEN 170
60 X1 = 1: H = DX; TA = 0
70 X2 = X1 + DX
80 B = 1/X1
90 A = B * H
100 TA = TA + A
110 IF TA > 1 THEN 150
120 PA = TA
130 X1 = X2
140 GOTO 70
150 T = X1 + (1 – PA)/(TA – PA) * DX
160 PRINT "T = ";T;" FOR A CHANGE OF X = ";DX
170 PRINT "TYPE 1 TO USE THE TRAPEZOID METHOD, 2
    OTHERWISE"
180 INPUT S
190 IF S <> 1 THEN 320
200 X1 = 1: TA = 0: H = DX
210 X2 = X1 + DX
220 B1 = 1/X1
230 B2 = 1/X2
240 A = .5 * (B1 + B2) * H
250 TA = TA + A
260 IF TA > 1 THEN 300
270 PA = TA
280 X1 = X2
290 GOTO 210
300 T = X1 + (1 – PA)/(TA – PA) * DX
310 PRINT "T = "; T; " FOR A CHANGE IN X = ";DX
320 PRINT "TYPE 1 IF YOU WANT TO CONTINUE, 2 TO STOP"
330 INPUT S
340 IF S = 1 THEN 10
350 END
```

By changing lines 110 and 260 you can calculate the value of *t* for any area. Of course, even the computer would not be able to calculate *t* for a large area (with divisions of 0.1 or less) due to the large number of calculations necessary. An interesting experiment might be to determine the largest area for which the computer could figure out a value of *t* in a given amount of time. If students have a good programming background, a challenge could be to rewrite the program so that it runs quicker (this version has many unnecessary calculations). This challenge could be made into a competition between students or groups of students. Or students could estimate the time it theoretically would take to calculate *t* for an area of 1000.

2.

The equation $e^{i\pi} + 1 = 0$ involves probably the five most important numbers in mathematics. The reason this relationship holds is similar to the reason $x^0 = 1$ (see the Extension and Enrichment section of "$F(x)$: Functions, Part 1" for the explanation); that is, $e^{i\pi} = -1$ is consistent with the laws of exponents that we do understand. Specifically, there is a theorem in analysis which states that if a definition of e^{a+bi} is consistent with the laws of algebra for $b = 0$ (real numbers), then it is the only such definition (*e.g.* e^{a+bi} should equal e^a if $b = 0$; $e^{a+b} = e^a e^b$). It was discovered that the definition which "works" is $e^{a+bi} = e^a (\cos b + i \sin b)$. If $a = 0$ and $b = \pi$, we have $e^{i\pi} = -1$.

3.

This area problem can be compared to the $G(x)$ problem if you notice that the function $A(t) =$ the area bounded by $y = 0$, $y = 1/x$, $x = 1$ and $x = t$ has the property that $A(ab) = A(a) + A(b)$.

Teaching Suggestions

1.

To help students get started and to be sure they understand the problem, present the following rough estimate of t for an area of one square unit:

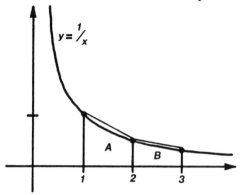

So as not to unnecessarily confuse the students, first review the formula for the area of a trapezoid and be sure they realize that both A and B are trapezoids. Then the area of $A = \frac{1}{2}(1 + \frac{1}{2})1 = \frac{3}{4}$ and the area of $B = \frac{1}{2}(\frac{1}{2} + \frac{1}{3})1 = \frac{5}{12}$. Through discussion, students should realize that t is between 2 and 3. Indicate to students that this is a very rough estimate of t and that their assignment is to determine the value of t for each of the three areas as best as they can.

2.

Let students know that the values of t are not rational; therefore, they are attempting to find the best rational approximation for t. You might add that it is unlikely (but not impossible!) that they will calculate t to the nearest thousandth and that their task is to do the best they can in the given time.

3.

Students should realize that they cannot figure out the value of t for an area of 1000 square units directly; i.e. the number is much too large (e^{1000}). It is not necessary, but you might want to imply that the problem is to discover a pattern. It is important not to suggest that more than two examples are needed to be confident of a pattern — an important focus of the discussion section is whether students generated more than two examples before concluding that there was a pattern.

4.

It is suggested that for this problem the students not be allowed to use computers or programmable calculators. The rationale for this restriction is that a major focus of the problem is the manipulative skills, especially the ability to organize and work with complicated calculations.

Problem 5   Calculating Area, Part 2

Student Solutions

1.

It should be noted that this nonroutine problem is equivalent to solving for t in the integral: $\int_1^t (1/x)\ dx = 1$, the solution being e (e^2 for 2 and e^{1000} for 1000). The most common student solution for an area of 1 square unit is to construct rectangles or trapezoids of height equal to 0.1 until just before an area of one is reached and then to use heights of 0.01. Many of these students do not realize that the error caused by using 0.1 is not "corrected" by using 0.01 at the end.

In contrast, one student consistently used heights of 1/64 to obtain a value of 2.719 ($e = 2.718...$). In addition the student discovered a pattern which allowed quick determination of the area of each trapezoid.

2.

Estimation can be used both properly and improperly for this problem. For example, one student used two trapezoids of height one and a third trapezoid of height 0.6 to obtain a final value of 2.6 for t. This may be a good initial estimate of t, but not a good final value. In contrast, many students get a similar estimate but use it to get a rough idea of the value of t and/or as a check on their later calculations.

3.

A key concept in the discussion of this problem is the proper use of induction. For example, many students will generalize an answer for the value of t for an area of 1000 based on just two bits of data, areas of 1 and 2. To emphasize the need for more examples before generalizing, have the students compare the following solutions from three students in one class (for consistency we will assume that each student found $t = 2.71$ for an area of 1 and $t = 7.37$ for an area of 2):

The first student "noticed that the difference between t for an area of 1 and t for an area of 2 is 4.66 which is about $4\frac{2}{3}$.... So in order to get the value of t for an area of 1000 you must multiply 999 times $4\frac{2}{3}$ and add it to 2.71. (Therefore,) the value of t for an area of 1000 is $2.71 + (4\frac{2}{3})(999)$."

The second student "found that when you square the t value for the area equal to 1 the answer is very close to the value of t when the area equals 2. I (the student) therefore concluded that when you put the value of t (when the area equals 1) to the power of the area, you will find the corresponding t value for the area equal to 1000 ($t = 2.71^{1000}$)."

The third student "decided to take the square root of t when the area under the curve equals 2.... That number is surprisingly close to 2.71, therefore I assert that t equals 2.71 raised to a power equivalent to the desired area under the curve. I tried this formula with areas of a half and 0 and it seemed to hold. (Therefore, I think) t would equal 2.71^{1000} when the area under the curve equals 1000."

After student discussion, the teacher should guide the students to the conclusion that the first and second student made the same reasoning error: generalizing with too few examples. From a problem-solving point of view the two students should receive the same grade *even though one student obtained the "correct" answer.* In contrast, the third student formed a hypothesis and tested it with two additional values.

4.

Another valuable focus for the discussion is the manipulative skills required for this problem (most solutions, though not all). Questions for discussion can include: Were mistakes made in the calculations? If so, what might have prevented the mistakes? How did students organize their work? How many students had to start over due to poor organization, etc.? Were any shortcuts used? How was work checked?

One student wrote: "After over an hour of work I realized that the first area was wrong, so I pulled my hair out and tried to stay calm and started again." Did any of your students have similar experiences?

5.

Some other techniques that students have used include:

- weighing, to get 2.72 for t for an area of 1. Based on solutions for areas of 2, 3, 4 and 5, the student concluded that the value of t for an area of 1000 was 2.72^{1000}.

- finding patterns in the area of the rectangles which allow them to solve for t. For example, one student discovered that for each area of approximately 0.697, there was a pattern; namely the first 0.697 went to $t = 2$, the second to $t = 4$, the third to $t = 8$, etc. Therefore, since $1000/0.697 = 1434.72$, the value of t for an area of 1000 should equal $2^{1434.72}$ which is approximately 2.703^{1000}.

- developing formulas for the areas to be calculated which allow completion of the problem in a systematic, concise manner.

Problem 6

Infinity, Part 1

Compare the size of the following sets:

{1, 2, 3, 4}	{all functions from (0,1) onto (0,1)}
{a, b, c, d}	{all real numbers between 0 and 1}
{1, 2, 3, 4, 5}	{all real numbers between −1 and 1}
{1, 2, 3,....}	{all real numbers between 0 and 1000}
{2, 4, 6,....}	{all rational numbers between 0 and 1}
{−5, −4, −3,....}	{all real numbers between 0 and ∞}
{1000, 2000, 3000,....}	{the grains of sand in the Sahara Desert to a depth of 10 feet}

Explain your rationale for labelling one set as being larger than another set. Your theory should not allow for any more than five sets being the same size. Also, at least two sets (two or more sets in each set) of infinite sets should be the same size.

Hint: How did you compare the first three sets? *Think about it!!* Can this method of comparison be extended....

Enrichment and Extension

1.

In addition to being an excellent enrichment topic, Cantor's theory of transfinite numbers is necessary to process "Infinity, Part 1" and to prepare for "Infinity, Part 2." Cantor considered two sets to be of the same (or equal) cardinality if there exists at least one one-to-one correspondence between the two sets. By this definition $\{1, 2, 3,....\}$ and $\{2, 4, 6,....\}$ are of equal cardinality because the 1-1 correspondence of $n \leftrightarrow 2n$ can be established. Similarly, $\{-5, -4, -3,....\}$ can be shown to have the same cardinality as $\{1, 2, 3,....\}$ using the map $n \leftrightarrow n - 5$, and $\{1000, 2000, 3000,....\}$ can be shown to have the same cardinality as $\{1, 2, 3,....\}$ using the map $n \leftrightarrow 1000n$.

By a more complicated proof, the set of all rational numbers can be shown to be in 1-1 correspondence with $\{1, 2, 3,....\}$. An outline of the proof is as follows: Let $f(n)$ be the function defined for non-zero integers so that $f(2n) = -n$ and $f(2n-1) = n$. Thus, $f(1) = 1, f(2) = -1, f(3) = 2, f(4) = -2$, *etc.* Each natural number greater than 1 has a prime factorization; let the prime factorization of $m = p_1^{a_1} p_2^{a_2} p_3^{a_3} \cdots p_k^{a_k}$ where $p_1 < p_2 < \cdots < p_k$. Then we can define a function $g(x)$ over the natural numbers such that for $m > 1$,

$$g(m) = p_1^{f(g_1)} p_2^{f(g_2)} p_3^{f(g_3)} \cdots p_k^{f(g_k)} \text{ and } g(1) = 1.$$

It is the case that $g(x)$ is a one-to-one function from the natural numbers to the positive rational numbers (some examples: $g(6) = 6, g(12) = 3/2$, $g(24) = 12, g(30) = 30, g(50) = 2/5, g(60) = 15/2$, and $g(17,640) = 20/21$). QED.

One of Cantor's greatest accomplishments was to demonstrate that the set of all real numbers between 0 and 1 has larger cardinality than the set $\{1, 2, 3,....\}$. To establish that set A is larger than set B you need to establish that there is a 1-1 map between B and a subset of A, but no such map between A and a subset of B. It is easy to demonstrate a 1-1 map between $\{1, 2, 3,....\}$ and a subset of $\{$all real numbers between 0 and 1$\}$ *e.g.* the map $n \leftrightarrow 1/n$. To

demonstrate the second condition is more difficult. Cantor did this by an indirect proof, assuming that the two sets were the same size and reaching a contradiction. He argued that if they were the same cardinality then by definition there would be at least one 1-1 correspondence between the two sets which could be represented as follows:

$$1 \leftrightarrow 0.a_{11} \, a_{12} \, a_{13} \, a_{14} \cdots$$
$$2 \leftrightarrow 0.a_{21} \, a_{22} \, a_{23} \cdots$$
$$3 \leftrightarrow 0.a_{31} \, a_{32} \, a_{33} \cdots$$

$$n \leftrightarrow 0.a_{n1} \, a_{n2} \, a_{n3} \cdots a_{nn} \cdots$$

where a_{ij} is the jth digit in ith number. Cantor showed that there was at least one real number between 0 and 1 not on the list. He constructed such a number by choosing the nth digit of the decimal expansion of the nth number to be any digit *but* a_{nn}. This number is not on the list because at least one digit of the new number differs from each number on the list; *e.g.* for each n, the new number differs from the number corresponding to n in at least one place, the nth digit. Contradiction. QED.

It should be noted that technically the proof needs to take into account the fact that there is not a unique representation for each real number, *e.g.* $0.4000....$ $= 0.3999.....$

The transfinite number c, equal to the number of real numbers between 0 and 1, can be thought of as the number of elements in the powerset of $\{1, 2, 3,....\}$. It can be shown that f which equals the number of elements in the set of all functions from $(0,1)$ onto $(0,1)$ has larger cardinality than c.

Teaching Suggestions

1.

Be sure the students understand that the restriction that no more than five sets can be the same size implies that it is not acceptable to label all the infinite sets as being equal in size. The restriction that at least two sets of infinite sets should be the same size implies that it is not acceptable to label the infinite sets from smallest to largest with no infinite sets being of equal size. The purpose of these restrictions is to force the students to develop a nontrivial ordering system.

2.

Encourage the students to consider the hint; that is, if they can figure out why they are able to compare the size of the sets $\{1,2,3,4\}$, $\{a,b,c,d\}$ and $\{1,2,3,4,5\}$ easily, perhaps they can extend that reasoning to the infinite sets.

3.

The set of all functions from $(0, 1)$ onto $(0, 1)$ needs to be clarified. It may be helpful to use the identity function $I(x) = x$ (restricted to $(0, 1)$) as an example of such a function. You might point out that like the irrational and transcendental numbers, almost all the functions in this set are rarely, if ever, used.

Additional examples will give the students a sense of the size of the set. For example, the function $F(x) = x$ for all x in $(0, 1)$ except x equals $1/2$ and $1/3$, for which $F(1/2) = 1/3$ and $F(1/3) = 1/2$. Explain to the students that this function is the same as $I(x)$ except that the values for $1/2$ and $1/3$ have been exchanged. Make it clear to students that as far as you know this is a completely useless function but it is an element in the set. To check for understanding, ask students to construct additional examples. They should be able to see that they can exchange any two values. Through discussion and questioning, students should realize that one could also exchange any three elements (*e.g.*, $F(a) = b$, $F(b) = c$, and $F(c) = a$), four elements, *etc.*

4.

In most classes, at least some students may need clarification concerning the definition of real numbers and the definition of rational numbers.

5.

The two lessons on infinity generate the strongest reaction on the part of the students. A reaction something like "Are you crazy? Asking us to compare infinite sets? Infinity is infinity, plus we don't understand any infinite sets." It might help to encourage students to work in pairs for this problem and to check student progress after the first week, perhaps even scheduling individual conferences.

6.

Optional. Many students think it's obvious that there are twice as many elements in $\{1, 2, 3, 4,....\}$ as in $\{2, 4, 6, 8,....\}$, arguing that the second set is contained in the first set and accounts for only half the elements. You can help them see the invalidity of this "proof" by arguing that if you divide each element of a set by the same number, you are not changing the number of elements in the set. For example, $\{4, 8, 12\}$ has the same number of elements as $\{4/2, 8/2, 12/2\}$. Once students agree with this point, you can divide each element in $\{2, 4, 6, 8,....\}$ by 4, resulting in the set $\{2/4, 4/4, 6/4, 8/4,....\}$ or $\{1/2, 1, 3/2, 2,....\}$, which by your previous argument must have the same number of elements as $\{2, 4, 6, 8,....\}$. Now comparing $\{1, 2, 3, 4,....\}$ and $\{1/2, 1, 3/2, 2,....\}$, you can argue (with a straight face!) that there are *obviously* twice as many elements in the second set since the first set is contained in the second set and accounts for only half the elements. Therefore, there are twice as many elements in $\{2, 4, 6, 8,....\}$ as in $\{1, 2, 3, 4,....\}$! At this point, you can close by concluding that you need to be careful in your arguments for discriminating among the sizes of infinite sets. Including this argument in your directions is not recommended if you feel the students are already overly concerned/worried by the difficulty of the assignment.

Student Solutions

1.

According to Cantor's Theory of Transfinite Numbers, the infinite sets for the assignment are ordered in the following way: the sets {1, 2, 3,....}, {2, 4, 6,....}, {−5, −4, −3,....}, {1000, 2000, 3000,....} and {all rational numbers between 0 and 1} have the same cardinality and are the smallest of the infinite sets; next in cardinality are the sets {all real numbers between 0 and 1}, {all real numbers between −1 and 1}, {all real numbers between 0 and 1000} and {all real numbers between 0 and ∞}; and finally the set of largest cardinality is {all functions from (0, 1) onto (0, 1)}. A summary of Cantor's Theory is outlined in the Enrichment and Extension section and probably should be included in the discussion, not only as part of this lesson, but also in preparation for "Infinity, Part 2."

2.

The most common line of reasoning that students use for this assignment is that if set *A* is contained in set *B*, then set *A* is smaller than set *B*. These students run into two major difficulties: satisfying the requirement that at least two sets of infinite sets must be the same size, and comparing sets where neither is contained in the other.

3.

Here are a few examples of the type of "sloppy" reasoning which is common in student solutions for this problem:

- One student had difficulty finding two sets of infinite sets of equal cardinality and finally equated {all functions from (0, 1) onto (0, 1)} with {all reals between 0 and ∞} because the size of both is "unimaginable and can't be compared to any of the others" and equated {all rationals between 0 and 1} and {1, 2, 3,....} "because the first set is simply 1 over the second set" (incorrectly calling 1/1, 1/2, 1/3,.... all the rationals between 0 and 1).

- One student noted that for the intervals "there's a definite beginning and end to them" and, therefore, they are smaller than sets such as {1, 2, 3,....} for which "there is no end."

- One student grouped the three finite intervals "because they all start with an element making a one-to-one correspondence at the beginning. It doesn't matter what the final elements in the sets are because it will never be reached. This is because there's an infinite number of real numbers between any two numbers. Because of this there will always be a one-to-one correspondence with the elements in these sets and the final element has no meaning."

- One student stated that "{2, 4, 6,....} and {all real numbers between 0 and 1} are of equal size, in my mind, because possibly a relationship could be established and each would seem to be one-half as large as the next sets ({1, 2, 3,....} and {all real numbers between −1 and 1})." Notice that the student has not given a reason why the sets determined to be of equal size are equal.

When compared to the other problems, the overall quality of solutions for the two problems on infinity may appear to be of lower quality; therefore, you might want to adjust your expectations accordingly.

4.

Here are a few examples of solutions, or partial solutions, judged to be of better quality than the preceding examples:

- One student argued that {all rationals between 0 and 1} was larger than {1, 2, 3,....} because the first set consisted of all the combinations of two whole numbers (smaller number as numerator) and supported his argument by comparing the size of {1, 2, 3, 4} and {1/1, 1/2, 1/3, 1/4, 2/3, 3/4}.

- One student carefully developed the concept of density as a tool to order the sets. According to the student, the density of two sets is the same if they have the same number of elements per equal interval. For example, the pair of sets {–5, –4, –3,....} and {1, 2, 3,....} were determined to have the same size and the three sets {reals between 0 and 1}, {reals between –1 and 1} and {reals between 0 and ∞} were determined to consist of sets of the same size.

- One student argued that the denumerable sets were all equal in cardinality because you were able to establish one-to-one correspondences between them, which were included in the solution.

- One student seemed to sense the basic difference between the denumerable sets and the intervals. The student commented: "In a sense it [the set of rationals between 0 and 1] is like a set of integers. Each fraction can be accounted for or counted by whole positive numbers. This set simply goes to infinity like the previous four sets [denumerable sets]. Maybe a way to say it is that the infinity in these sets is the smallest kind of infinity. You can perceive it as one dimensional." Further, the student described the sets of intervals: "Between every single pair of numbers in this set, there is another set of infinite numbers; and between two of those numbers is another set of numbers....Because this infinity is so huge starting at 0 or whatever makes no difference in size. The sets of this group belong to a higher order of infinity. It is almost like a two-dimensional infinity."

- One student used the concept of "double infinite set" to label {all real numbers between 0 and ∞} and {all functions from (0, 1) onto (0, 1)} as larger than the other infinite sets. For example, for the set {all reals between 0 and ∞} the student argued "first all the reals is an infinite set and between 0 and infinity is another infinite set, making this a double infinite set."

Problem 7

Grading $G(x)$

Your assignment is to develop a short yet valid method for grading the $G(x)$ assignment. Your work will be graded on two criteria: (1) is the method valid? That is, would the results of your method be similar to the results of a panel of experts? and (2) how long does your method take to grade a group of papers? The less time required to grade the papers the better.

The information you were given for the $G(x)$ problem and the actual values of $G(x)$ are given below:

I. You are given the following information about a function $G(x)$:

A. For all $a, b > 0$, $G(ab) = G(a) + G(b)$

B. $G(8) = 1.660$

  $G(14) = 2.107$

  $G(6) = 1.430$

  As best you can, complete the following values.

$G(1) = 0$	$G(11) = 1.914$
$G(2) = .553$	$G(12) = 1.984$
$G(3) = .877$	$G(13) = 2.047$
$G(4) = 1.107$	$G(14) = 2.107$
$G(5) = 1.285$	$G(15) = 2.162$
$G(6) = 1.430$	$G(16) = 2.213$
$G(7) = 1.553$	$G(17) = 2.262$
$G(8) = 1.660$	$G(18) = 2.307$
$G(9) = 1.754$	$G(19) = 2.350$
$G(10) = 1.838$	$G(20) = 2.391$

Problem 7 Grading G(x)

Extension and Enrichment

1.

Below are the directions for a nonroutine problem which can provide good enrichment for this problem:

Develop a method for assigning a number to indicate the degree of similarity between two rank orders of the same ten items. The range of the numbers in your method and the meaning of the numbers should be specified. At least three types of similarity should be taken into account: (1) rank orders which are very similar, (2) rank orders which are very dissimilar and (3) rank orders which seem not to be related. For example, you could develop a method which assigns a number between 20 and 40 to two rank orders of ten items. A score close to 40 would indicate very similar rank orders, 20, very dissimilar, and 30, no relationship.

Here are some sample data. We'll assume that the first column is one person's rank order of ten movies from best to worst. The next three sections demonstrate rank orders that could be considered similar, dissimilar and not related. Your method should be able to assign different numbers (perhaps a few identical scores) to each rank order in a consistent and sound way.

2.

After work on the nonroutine problem, the rank correlation formula can be introduced:

$$r = 1 - \frac{6 \Sigma d^2}{n(n-1)}$$

where d = the difference between two ratings, n = the number of items ranked. The formula, or perhaps techniques that the students developed, can be used to evaluate the validity of the procedures for marking $G(x)$; *e.g.* compare the teacher's rank order of solutions with student rank orders.

Rank	Similar			Dissimilar				No Relationship		
1	2	2	1	10	1	9	10	5	7	6
2	3	1	2	9	10	8	9	8	5	2
3	4	4	3	8	9	7	8	2	1	8
4	5	3	6	7	8	6	7	6	9	4
5	6	6	5	6	7	5	4	3	3	10
6	7	5	4	5	6	4	5	10	6	1
7	8	8	7	4	5	3	6	7	2	5
8	9	7	8	3	4	2	3	1	10	9
9	10	10	9	2	3	1	2	4	8	3
10	1	9	10	1	2	10	1	9	4	7

Teaching Suggestions

1.

To introduce the problem, tell the students that this problem will be based on the results of the $G(x)$ problem, particularly how to mark the papers. Discuss the fact that, theoretically, an excellent way to mark the papers would be to have a panel of experts read all the papers and then score each paper. Point out that this method would be impractical if there were a large number of papers to mark. Suggest the following alternative: for each student, calculate $S = \Sigma \,|$[actual value of $G(x)$] − [student's value of $G(x)$]$|$, then give grades according to the value of S. The lower the value of S, the higher the grade. Through further discussion, students should realize that this method of marking would be reasonably accurate and certainly quicker than using a panel of experts; however, this method would still be lengthy for marking 1000 papers.

2.

The assignment is to develop a method which would yield results similar to a panel of experts (*i.e.* a valid measure). Emphasize that they will be graded on the validity of their method of marking and the length of time required to grade a paper using their method (the shorter, the better).

3.

Your assessment of your students should determine how much information you give to help them get started on the problem. Generally, two types of information can be useful; first, information concerning the $G(x)$ problem such as: (a) the values of $G(x)$ for x equalling 2, 3, 4, 6, 7, 8, 9, 12, 14, 16 and 18 can be determined directly and fairly easily from the given values; (b) $G(1)$ can be determined from the given information but is not as straightforward a calculation; (c) the values of $G(x)$ for x equalling 5, 10, 11, 13, 15, 17, 19 and 20 are more difficult to determine — in addition, 10 and 11, and 19 and 20 are adjacent difficult values; and the value for x equals 5 affects three other values (10, 15 and 20); and (d) the larger the value of x, the more accurate averaging becomes for determining unknown values; *e.g.* $G(5) = 1.285$ while $[G(4) + G(6)]/2 = 1.269$, whereas $G(19) = 2.350$ while $[G(18) + G(20)]/2 = 2.349$, only .001 off.

Secondly, you might want to give them some sample student solutions as data. Three sample solutions are attached as a potential handout; however, a better option might be to give the students copies of the solutions of their class to the $G(x)$ problem. That way the data is relevant and more real. If you pick this last option, you might want to use your grades as a method of comparison (you being the panel of experts!) and even consider withholding your grades for the $G(x)$ problem until after they have completed this assignment.

Problem 7   Grading G(x)

Data Sample

Below you are given solutions from three different students:

I.

The first student calculated the values for all the easily determined values and $G(1)$ directly from the given information. Then the student determined the remaining values using excellent approximations based on easily determined values of $G(x)$ for large values of x.

$G(1) = 0$	$G(11) = 1.914$
$G(2) = .553$	$G(12) = 1.983$
$G(3) = .877$	$G(13) = 2.048$
$G(4) = 1.106$	$G(14) = 2.107$
$G(5) = 1.285$	$G(15) = 2.162$
$G(6) = 1.430$	$G(16) = 2.212$
$G(7) = 1.553$	$G(17) = 2.262$
$G(8) = 1.660$	$G(18) = 2.307$
$G(9) = 1.754$	$G(19) = 2.351$
$G(10) = 1.837$	$G(20) = 2.391$

II.

The second student discovered that $G(x) = k[\log x]$, but calculated $k = 1.84$ versus the more accurate 1.838. Also, the student did not check the values obtained this way with the values obtained directly from the given values.

$G(1) = 0$	$G(11) = 1.916$
$G(2) = .554$	$G(12) = 1.986$
$G(3) = .878$	$G(13) = 2.049$
$G(4) = 1.108$	$G(14) = 2.107$
$G(5) = 1.286$	$G(15) = 2.164$
$G(6) = 1.430$	$G(16) = 2.214$
$G(7) = 1.555$	$G(17) = 2.264$
$G(8) = 1.660$	$G(18) = 2.310$
$G(9) = 1.756$	$G(19) = 2.353$
$G(10) = 1.840$	$G(20) = 2.394$

III.

The third student calculated the values for all the easily determined values and $G(1)$ directly from the given information. Then the student determined the remaining values using graphing techniques.

$G(1) = 0$	$G(11) = 1.915$
$G(2) = .553$	$G(12) = 1.983$
$G(3) = .877$	$G(13) = 2.043$
$G(4) = 1.107$	$G(14) = 2.107$
$G(5) = 1.282$	$G(15) = 2.159$
$G(6) = 1.430$	$G(16) = 2.213$
$G(7) = 1.553$	$G(17) = 2.269$
$G(8) = 1.660$	$G(18) = 2.307$
$G(9) = 1.754$	$G(19) = 2.367$
$G(10) = 1.835$	$G(20) = 2.389$

Student Solutions

1.

Most student solutions revolve about reducing the number of calculations by looking at just a few key values. For example, you could take (a) one value from $G(2)$, $G(3)$, or $G(7)$; (b) the value of $G(1)$; (c) one value from $G(4)$, $G(6)$, $G(8)$, $G(9)$, $G(12)$, $G(14)$, $G(16)$ or $G(18)$; and (d) perhaps two values from $G(5)$, $G(10)$, $G(11)$, $G(13)$, $G(15)$, $G(17)$, $G(19)$ or $G(20)$. Calculating $S = \Sigma|G(x)$ – student's value of $G(x)|$ for these five numbers would be an example of one possible solution.

2.

Another alternative is to pick a similar set of numbers but to assign different weights to some values; for example, if $G(20)$ was considered the most significant number then $|G(20)$ – the student's $G(20)|$ could be multiplied by three to increase its effect on the grade. One student tripled the weight of $G(11)$ because "it is the lowest prime value that cannot be determined [directly] that is next to another value which cannot be determined $[G(10)]$."

3.

Some students develop methods that can be carried out by inspection of a few key values. For example, one student developed a system based on the values of $G(1)$, $G(5)$ and $G(16)$, reasoning that "If a student determines that $G(1)$ is 0, he thoroughly understands the relationship $G(ab) = G(a) + G(b)$.... If you determine $G(16)$ to be 2.213 you have had to figure out that the function is log base 3.5, because when you determine the values arithmetically... you find that $G(16) = 2.214$.... Averaging $G(14)$ and $G(16)$ gives a value for $G(15)$ that is only off by 0.001. But when you average $G(4)$ and $G(6)$ the value for $G(5)$ is off by 0.016."

Based on these observations the student developed the following grading system with a 4.0 scale: (a) 1 point if $G(1) = 0$, (b) $\frac{1}{2}$ point if $G(5) = 1.285$ and $G(16) = 2.213$, (c) 1 possible point: 1 point if $G(16)$ is between 2.212 and 2.214 or $\frac{1}{2}$ point if $G(16)$ is between 2.208 and 2.218, and (d) $1\frac{1}{2}$ possible points:

$1\frac{1}{2}$ points if $G(5) \in [1.275, 1.295]$.
1 point if $G(5) \in [1.270, 1.275) \cup (1.295, 1.300]$.
$\frac{1}{2}$ point if $G(5) \in [1.269, 1.270) \cup (1.300, 1.301]$.
$\frac{1}{4}$ point if $G(5) \in [1.26, 1.269) \cup (1.301, 1.310]$.

4.

One suggestion for the discussion is to have the students try out their methods on actual or made up solutions and compare the results to the teacher's assessment or a longer method of calculation (see the enrichment section). It could be beneficial for students to try out other students' methods either in pairs or small groups. One focus could be the clarity of the methods, an important skill in developing algorithms.

Problem 8

Connecting *n* Random Points

The problem is to develop a procedure to connect n random points so that the total distance is minimized. You are required to write a procedure that another person can follow. In addition, if two people follow the procedure, the results should be the same. The procedure will be tested by other students on a sheet ($8\frac{1}{2} \times 11$ paper) of random points (at least fifteen points) provided by the teacher.

Extension and Enrichment

1.

There is a nice indirect proof that the shortest length must include the shortest connection between any two points: Assume that point *A* is the closest point to *B*, but is not connected to *B*. For example, let *A* be connected to point *C* instead:

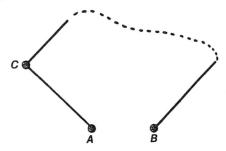

Since *AB* must be shorter than *AC*, if you erase *AC* and draw *AB* instead the total length must be less. Contradiction. QED.

2.

The problem of determining a general procedure for constructing the shortest closed path for connecting *n* points remains unsolved. There is a theorem that the shortest path cannot include two lines that cross. Students could try to develop procedures for this type of path and compare their results on two or three sample sets of points.

3.

Although not directly mentioned, an assumption of the problem is that only the given dots are connected. Without that assumption there are cases where adding a point (or more) can decrease the total distance. For example, the four vertices of a square would be connected thusly by the described procedure:

However, if you add the center point of the square, you get the following better solution:

Notice that if the side of the square is $\sqrt{2}$ units, then the first solution would involve $3\sqrt{2}$ units (approximately 4.2 units) versus 4 units for the second solution.

Teaching Suggestions

1.

The following statement can be used to introduce the problem: Suppose the telephone company wants to develop a procedure that any (qualified) employee can follow which will tell the employee how to connect *n* telephone poles using the least amount of wire. Emphasize that if any two employees follow the procedure correctly they should end up connecting the poles the same way.

2.

Put the following diagram on the chalkboard:

3.

Ask the students what they believe the best way would be to connect these points (minimizing distance).

The best answer is:

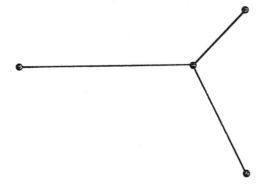

4.

The previous example should alert the students to the difficulty of the problem. The assignment is to develop a procedure for connecting *n* points randomly placed by the teacher on an $8\frac{1}{2} \times 11''$ piece of paper so as to minimize the distance. Emphasize that any two students (or teachers) should be able to carry out the procedure and obtain the same result.

5.

Tell the students that their procedures will be used to connect a set of random points (at least fifteen) provided by the teacher. The total distance for each procedure will be compared to the results of other students' procedures.

Student Solutions

1.

Theoretically the best solution is to first connect each point to the point nearest it. This process results in a number of unconnected "links." Then connect links by connecting the two closest points of nearby links. Some students discover this method or equivalent methods, some of which take more time or include unnecessary steps.

2.

Other techniques students develop include (a) using spiralling patterns, (b) starting with a central point and drawing "radii" of connected points, (c) making use of angles; *e.g.,* using the fact that for a given base the legs of an isosceles triangle get smaller as the angle increases, (d) developing a method to select an initial point, then connecting to a second point, third point, *etc.,* (e) dividing the sheet into sections, usually quadrants, and then working in one section at a time, (f) avoiding or eliminating any polygons formed, (g) working with clusters, then connecting the clusters, and (h) working with naturally forming "vertical" or "horizontal" lines, then connecting the lines.

3.

After reviewing the papers, the following activity could be fruitful: (a) have students exchange procedures and identify any directions which are unclear, (b) give the procedures and questions back to the student who developed the procedure and have that student rewrite any unclear directions, (c) give each student the attached handout *and* another student's procedure with the following assignment: follow the procedure for the three boxes and measure the total distance in centimeters, and (d) discuss the results the next day. An additional option is to have each procedure tried by two other students, thereby checking whether the procedure satisfies the condition of generating a unique result.

The activity tends to focus the student's attention on a key skill for this problem: developing algorithms. They get direct, concrete feedback on the clarity and effectiveness of their algorithm and are exposed to a variety of other solutions.

4.

A sample page with three sets of random points is attached. The best solution for the two sets at the top of the sheet is easily determined by inspection; therefore, these two sets provide good feedback to the students on the effectiveness of their procedure for simple cases. The lower set is more complicated and can be used for comparing procedures.

Sample Set of Points

Problem 9

$R(x)$: Functions, Part 3

You are given the following information about the function $R(x)$:

1. $R(x) = R(x-1) + x$

2. $R(1) = 2$

I.

Determine the following values of $R(x)$:

$R(1) =$	$R(11) =$
$R(2) =$	$R(12) =$
$R(3) =$	$R(13) =$
$R(4) =$	$R(14) =$
$R(5) =$	$R(15) =$
$R(6) =$	$R(16) =$
$R(7) =$	$R(17) =$
$R(8) =$	$R(18) =$
$R(9) =$	$R(19) =$
$R(10) =$	$R(20) =$

II.

The following values of $R(x)$ are more difficult to determine. Some require you to discover a pattern. Some cannot be calculated with the given information and require you to extend the definition of $R(x)$. Determine the best values you can for the five values of the function below. If you extend the definition of $R(x)$, you are required to give a justification for the extension. Also, document any pattern you use.

$$R(-1000) =$$

$$R(2000) =$$

$$R(55\tfrac{1}{2}) =$$

$$R(7\tfrac{4}{5}) =$$

$$R(8.732) =$$

Extension and Enrichment

1.

One extension of this problem is to study recursive functions in a computer language such as PASCAL. One activity is to write a program which prints out values of *R(x)* and to see which values can or cannot be printed out (nonintegral cannot). This activity could lead to a study of discrete functions defined for only integral values.

2.

The Fibonacci sequence defined by $F(n+1) = F(n) + F(n-1)$ with $F(1) = 1$ and $F(2) = 1$ provides an excellent topic for enrichment. The article "Fibonacci Sequences" by Brother Alfred Brousseau in the NCTM's *Readings for Enrichment in Secondary School Mathematics* is an excellent source of information and problems on the Fibonacci sequence.

3.

A very interesting recursive function used by Archimedes to estimate pi is discussed in the Extension and Enrichment section of the problem on pi.

Teaching Suggestions

1.

Explain that the first part of the assignment, calculating $R(x)$ for $x = 1$ to 20, is straightforward. You might calculate $R(2)$, $R(3)$ and $R(4)$ as a class to be certain that students understand the recursive definition. Point out that the difficult part of the problem is Part II and that they can use the data from Part I to help figure out Part II.

2.

Students should be aware that figuring out a value such as $R(2000)$ by determining all the values from $R(1)$ to $R(2000)$ is not recommended. They should, however, look for patterns as they did for some of the other nonroutine problems. As in the $F(x)$ and $G(x)$ problems, it should be made clear that the function is not linear; therefore, the student cannot simply calculate, for example, $R(7\frac{4}{5})$ as $R(7) + \frac{4}{5}[R(8) - R(7)]$.

3.

It may be appropriate to discuss the difference between a discrete and a continuous function. It should be noted that $R(x)$ is a discrete function that is not defined for nonintegral values. Therefore, when students assign a value to $R(55\frac{1}{2})$, for example, they need to justify this extension of the function to a nonintegral value.

Student Solutions

1.

In this problem, the function $R(x) = .5x^2 + .5x + 1 = x(x+1)/2 + 1$ for integral values of x. Students that discover this pattern do so in a variety of ways. One student drew a graph with easily obtained values, hypothesized that the graph was a parabola, and solved for a, b and c in $R(x) = ax^2 + bx + c$ by substituting three pairs of values. Most students who determine $R(x)$ do not find the solution so directly. For example, one student first experimented with the two equations:

$$R(x) = R(x-1) + x$$

$$R(x+1) = R(x) + x + 1$$

to obtain:

$$R(x+1) - R(x) = R(x) - R(x-1) + 1$$

but the student concluded: "This makes perfect sense since each time x increases by 1, it increases the difference between two consecutive values by 1 too. But it doesn't help much." Then the student tried graphing and hypothesized it was a parabola, but first worked with the forms $y = x^2 + c$ and $y = x^2 + x + c$ before trying $y = \frac{1}{2}x^2 + \frac{1}{2}x + c$. This second student's solution is more typical of the "meandering" involved in most student solutions.

2.

Other students discover the pattern that $R(x)$ is one more than the sum of 1, 2, 3,...,x; that is, $R(x) = x(x+1)/2 + 1$.

3.

The most common solution by students who do not discover the equation is by graphing. Of course, the accuracy of this solution depends upon the manipulative skills of the student. How carefully are the points graphed? What scale is used? *etc.* For example, one student obtained a value of 35.3 (versus 35.32) for $R(7\frac{4}{5})$ by careful graphing, including a check by determining $R(1\frac{4}{5})$ by graphing and using the definition to get $R(7\frac{4}{5})$.

4.

As in the $F(x)$ and $G(x)$ problems, in addition to a focus on the three steps of problem solving, it is valuable to focus the discussion on manipulative skills and looking for patterns. A suggestion for this problem is to have the students discuss the solutions in small groups. For example, you could put the three steps of problem solving and the two additional problem-solving skills on the chalkboard and have the students find positive and negative examples from the group's solutions. Each group could report on their observations for a class discussion.

Problem 10

Infinity, Part 2

Simplify the following:

1. $\aleph_0 + \aleph_0 =$ 2. $5\aleph_0 =$ 3. $\aleph_0^2 =$

4. $c - \aleph_0 =$ 5. $\aleph_0 - \aleph_0 =$ 6. $\aleph_0 - 2 =$

7. $2^{\aleph_0} =$ 8. $1 + 4\aleph_0 =$ 9. $c + \aleph_0 =$

10. $2^c =$ 11. $2^{2^{\aleph_0}} =$

$\aleph_0$ equals the number of elements in the set $\{1, 2, 3,....\}$, c equals the number of real numbers between 0 and 1, and f equals the number of functions from (0,1) onto (0,1).

Provide documentation and support for your answers.

Extension and Enrichment

1.

What does it mean to say that the probability of picking a rational number from the set of real numbers between 0 and 1 is 0? One definition is that the probability of an event is 0 if, for any epsilon greater than 0, it can be demonstrated that the probability of the event is less than epsilon. If we can understand the significance of this definition of a probability of 0 we can begin to understand the difference in size between sets of cardinality $\aleph_0$ and sets of cardinality c. For example, the following events do *not* have a probability of 0: (a) a person randomly buries a small object on a beach on the Atlantic shore, the probability that you will find the object by randomly picking a spot to try, (b) two people randomly pick dates (including time of day to the nearest second) between 1,000,000 BC and the present, the probability that they will pick the same date and (c) each member of a class randomly picks a billion-digit number, the probability that each student will pick the same number. Finally, the Buddha gave an approximation of the probability of a nonhuman being reincarnated into a human: Suppose there is a blind turtle at the bottom of a great ocean. Somewhere on the surface of the ocean is a ring of wood floating on the waves, with the wind blowing it back and forth. The probability that if the turtle surfaces it will put his head through the ring of wood is the probability of a human birth. Again this probability is not zero.

2.

A good problem would be to have students approximate the probabilities of the previous events.

3.

This problem is an excellent opportunity to have the students compare the growth of 2^n and n^2. Remember, $\aleph_0^2 = \aleph_0$ but $2^{\aleph_0} = c$.

Teaching Suggestions

1.

Review the following ideas from the lesson "Infinity, Part 1": (a) two sets have the same cardinality if there exists at least one 1-1 correspondence between the two sets, (b) the sets $\{1, 2, 3,....\}$, $\{-5, -4, -3,....\}$, $\{1000, 2000, 3000,....\}$, $\{2, 4, 6,....\}$ and {all rational numbers between 0 and 1} have the same cardinality which we represent by $\aleph_0$, {c} the sets of all real numbers in the intervals $(0, 1)$, $(-1, 1)$, $(0, 1000)$ and $(0, \infty)$ all have the same cardinality which we represent by c, (d) the set of all functions from $(0, 1)$ onto $(0, 1)$ is the largest set discussed and the cardinality of this set is represented by f, (e) the number of subsets of a set of n elements is 2^n, and (f) c can be thought of as being equal to the number of subsets of a set with $\aleph_0$ elements.

2.

It is helpful to add the following four hints: first, the number of lattice points (a, b) such that a and b are positive integers is equal to $\aleph_0$ (this hint can be used to argue that $\aleph_0^2$ equals $\aleph_0$). Second, the probability of randomly picking a rational number from the set of reals is zero. There are a number of ways that you can add a little drama in presenting this hint. For example, you could ask students to guess the probability of picking a rational number from an imaginary bag of all the rationals and the irrationals.

Third, you can tell them the story of Hilbert's Hotel, a hotel with an infinite number of rooms numbered 1, 2, 3,.... in a line from left to right. One day all the rooms are occupied and a person asks for a room. The desk clerk replies that there is no problem and proceeds to announce to the guests: "Please move to the room to your right (*i.e.*, $n \rightarrow n + 1$)." Then he gives the new guest the room numbered one. Soon an infinite group of new guests arrives asking for rooms. Again the desk clerk replies that there is no problem. He announces: "All guests move to the room whose number is twice your present room number (*i.e.*, $n \rightarrow 2n$)." He then

assigns the odd numbered rooms to the new guests. Of course, you can embellish the story to add interest.

Finally, it helps to show students how sets can be created to represent some of the expressions. For example, since $\{1, 2, 3,....\}$ has $\aleph_0$ elements, then the set $\{-1, 0, 1,....\}$ has $2 + \aleph_0$ elements and the set $\{1, \frac{3}{2}, 2, \frac{5}{2},....\}$ has $2\aleph_0$ elements. A major mistake students make is to equate performing arithmetic operations on each member of a set with changing the number of elements in a set. For example, students will argue that if $\{1, 2, 3,....\}$ has $\aleph_0$ elements, then $\{2 \times 1, 2 \times 2, 2 \times 3,....\}$ must have $2\aleph_0$ elements, or $\{1^2, 2^2, 3^2, ...\}$ must have $\aleph_0^2$ elements. It is recommended that students understand this type of error before attempting the problems.

4.

It should be noted that number 5 (simplify $\aleph_0 - \aleph_0$) is the most difficult problem. This expression is undefined because one can argue that it equals 0, any finite number, of $\aleph_0$ see Solutions). Students are not generally aware of the difficulty of this problem so you may like to alert them and ask them to check their answer with you if they believe they have a correct answer.

5.

One effective method of evaluation is the following: Let the students know that they will get 0 to 2 points for each expression. Two points means that the simplification is correct and a good supporting reason is given; one point means that the simplification is correct but the supporting reason is weak or invalid; and 0 points means the simplification is incorrect. Additionally, you might give either 1 or $\frac{1}{2}$ points if the simplification is incorrect but the supporting argument is good.

6.

It should be noted that technically necessary definitions have not been given here for some of the notation used such as $2 + \aleph_0$, $\aleph_0^2$, and $2^{\aleph_0}$. These definitions are not necessary for the objectives of this set of problems and, in fact, could unnecessarily complicate the problem for students.

Student Solutions

1.

For each problem I will give the solution according to Cantor's theory and at least one example of an explanation. (1) $\aleph_0 + \aleph_0 = \aleph_0$: $\{1, 3, 5, 7,....\}$ and $\{2, 4, 6, 8,....\}$ both have $\aleph_0$ elements or, together, $\aleph_0 + \aleph_0$, but the combined set $\{1, 2, 3,....\}$ has $\aleph_0$ elements. (2) $5\aleph_0 = \aleph_0$: $\{\frac{1}{5}, \frac{2}{5}, \frac{3}{5}, \frac{4}{5}, 1, \frac{6}{5},....\}$ can represent $5\aleph_0$ elements, but there is a 1-1 correspondence between that set and $\{1, 2, 3,....\}$; *i.e.*, n $\leftrightarrow$ 5n. (3) $\aleph_0^2 = \aleph_0$. The number of lattice points can represent a set with $\aleph_0^2$ elements; recall the number of lattice points is equal to $\aleph_0$. (4) and (9) $c - \aleph_0$ and $c + \aleph_0 = c$: You can argue that this is like removing a bucket of water from the ocean or use the fact that the probability of selecting a rational number from the reals is 0. (5) $\aleph_0 - \aleph_0$ is undefined: as previously remarked, this can be shown to be equal to 0, or any finite number. For example, the following two sets have $\aleph_0$ elements: $\{1, 2, 3,....\}$ and $\{5, 6, 7, 8,....\}$; $\aleph_0 - \aleph_0$ can represent the number of elements in the set formed by removing the elements of the first set from the second set; that is, 4. Further, $\aleph_0 - \aleph_0$ can be shown to equal $\aleph_0$ because both $\{1, 2, 3,....\}$ and $\{2, 4, 6,....\}$ have $\aleph_0$ elements; therefore, the set $\{1, 3, 5,....\}$ formed by removing the even numbers from the whole numbers represents $\aleph_0 - \aleph_0$ but also obviously has $\aleph_0$ elements. (6) and (8) $\aleph_0 - 2$ and $1 + 4\aleph_0$ both equal $\aleph_0$ by comparing the following sets to $\{1, 2, 3,....\}$: $\{3, 4, 5,...\}$ and $\{0, \frac{1}{4}, \frac{2}{4}, \frac{3}{4}, 1, \frac{5}{4},....\}$. (7), (10), and (11) $2^{\aleph_0} = c$, $2^c = f$, $2^{2^{\aleph_0}} = f$: These can be argued by using the fact that 2^n equals the number of subsets in a set of n elements.

2.

As was implied in "Infinity, Part 1," the quality of answers that students give for these problems is generally poor. In discussing this lesson, you might focus on how the students handled the problems which they found more difficult. More specifically, you might ask the students to generate examples of instances where they persisted successfully (an indicator of this would be an "aha" experience) and instances where they did not persist (an indicator of this would be "settling" for an answer that you know is incorrect).

3.

Many of the correct solutions that students give follow directly from the hints. A few more interesting student solutions follow.

- One student illustrated equalities visually; for example, $5\aleph_0 = \aleph_0$ was illustrated thusly:

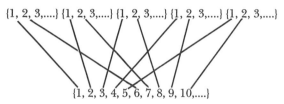

- One student argued that $c - \aleph_0 = c$ because "subtracting such a small number like the rationals from the whole original set is meaningless. It is comparable to $\aleph_0 - 5 = \aleph_0$. Since $\aleph_0$ is the number of rationals which is so tiny compared with the irrationals, subtracting it from c ... has little effect. Just as 5 has no effect on $\aleph_0$."

- One student illustrated $5\aleph_0 = \aleph_0$ by the following:

$$\aleph_0 = \{1, 6, 11,....\}$$
$$\aleph_0 = \{2, 7, 12,....\}$$
$$\aleph_0 = \{3, 8, 13,....\}$$
$$\aleph_0 = \{4, 9, 14,....\}$$
$$+ \ \aleph_0 = \{5, 10, 15,....\}$$
$$\overline{\qquad\qquad\qquad\qquad}$$
$$5\aleph_0 = \{1, 2, 3, 4,....\} = \aleph_0$$